BEAKS AND BILLS

Mel Higginson

Rourke
Publishing LLC
Vero Beach, Florida 32964

www.rourkepublishing.com

PHOTO CREDITS: All photos © Lynn M. Stone.

Title page: The bright colors of toucan's beak attract mates.

Editor: Robert Stengard-Olliges

Cover design by Nicola Stratford.

Library of Congress Cataloging-in-Publication Data

Higginson, Mel
Beaks and bills / Mel Higginson
p. cm. -- (Let's look at animals)
Includes index.
ISBN 1-60044-169-6 (Hardcover)
ISBN 1-59515-527-9 (Softcover)
1. Bill (Anatomy)--Juvenile literature. I. Title. II. Series: Higginson, Mel
M. Let's look at animals.
QL697.S77 2007
573.3'5518--dc22

2006012745

Printed in the USA

CG/CG

Rourke Publishing

www.rourkepublishing.com – sales@rourkepublishing.com
Post Office Box 3328, Vero Beach, FL 32964

Table of Contents

Beaks

Bird beaks grow outward from their mouths. The upper mandible and the lower **mandible** make up the beak. Beaks are also called bills.

Beaks grow from skin. Beaks grow much harder and tougher than normal skin.

This vulture's beak is hard and sharp. It cuts through tough skin.

Beaks are Tools

Bird beaks, or bills, are like tools. They have many uses. They have many sizes and shapes.

A woodpecker's beak is a hammer and pick.

A pelican's bill is a fish net. A pelican uses its pouch to catch fish.

Beaks for Eating

The eagle's beak is a meat hook. It is also the bird's pliers and scissors.

A bird's beak is most useful at dinnertime. The beak brings food into the body.

Beaks Catch Dinner

Some beaks even find dinner through their sense of touch! One such beak is the spoonbill's. The beak can feel little animal **prey**.

The **ibis's** long, curved bill pokes into water and mud. The bill's mandibles feel little animals like shrimp.

The flamingo's bill **filters** water. A filter traps food bits while water passes through the bill.

Many beaks catch dinner. Herons stand still and watch. They use their sharp beaks to grab fish.

Beaks are Useful

Beaks are useful to clean and straighten feathers. This is called **preening**. Birds use their beaks to drive other animals away.

A white pelican takes a time-out from babysitting to preen.

Some beaks change color. They probably help a bird **attract** a mate.

The puffin's bill has bright colors only during the nesting season.

A wood stork's beak carries sticks for the nest. Some penguins carry pebbles in their beaks for nests.

Some birds touch beaks. That seems to help make them get along together.

Glossary

attract (uh TRAKT) – to draw closer

filter (FIL tur) – to take small bits of food out of water through a screen of some kind

ibis (eye BIISS) – a wading bird with a long, down-curved beak

mandible (MAN duh bul) – the upper and lower parts of a bird's beak

preening (PREE ning) – the cleaning of feathers with beak or claws

prey (PRAY) – an animal that is food for another animal

Index

FURTHER READING

Collard, Sneed B. *Beaks!* Charlesbridge Publishing, 2002.
Pascoe, Elain. *Birds Use Their Beaks*. Gareth Stevens Audio, 2002.

WEBSITES TO VISIT

http://mdc.mo.gov/nathis/birds/peekbeak

ABOUT THE AUTHOR

Mel Higginson writes children's nonfiction and poetry. This is Mel's first year writing for Rourke Publishing. Mel lives with his family just outside of Tucson, Arizona